D0408277

MY BRIEF HISTORY

MY BRIEF HISTORY

STEPHEN HAWKING

BANTAM BOOKS NEW YORK

Published in the United States by Bantam Books, an imprint of The Random House Publishing Group, a division of Random House LLC, a Penguin Random House Company, New York.

BANTAM BOOKS and the HOUSE colophon are registered trademarks of Random House LLC.

Illustration credits appear on page 127.

Published simultaneously in the United Kingdom by Bantam Press, part of Transworld Publishers, a member of The Random House Group, London.

Portions of this work originally appeared in different form as part of lectures given by the author throughout the years.

LIBRARY OF CONGRESS CATALOGING-IN-PUBLICATION DATA
Hawking, Stephen.
My brief history / Stephen Hawking.
pages   cm
ISBN 978-0-345-53528-3 — ISBN 978-0-345-53913-7 (eBook)
1. Hawking, Stephen, 1942–   2. Physicists—Great Britain—Biography.
3. Cosmology.   4. Black holes (Astronomy)   I. Title.
QC16.H33A3 2013
530.092—dc23
[B]
2013027938

Printed in the United States of America on acid-free paper

www.bantamdell.com

9 8 7 6 5 4 3 2 1

First Edition

Book design by Liz Cosgrove

For William, George, and Rose

CONTENTS

MY BRIEF HISTORY

1

CHILDHOOD

M Y FATHER, FRANK, CAME FROM A LINE OF TENANT
farmers in Yorkshire, England. His grandfather—
my great-grandfather John Hawking—had been a wealthy
farmer, but he had bought too many farms and had gone
bankrupt in the agricultural depression at the beginning
of this century. His son Robert—my grandfather—tried
to help his father but went bankrupt himself. Fortunately, Robert's wife owned a house in Boroughbridge
in which she ran a school, and this brought in a small
amount of income. They thus managed to send their son
to Oxford, where he studied medicine.

My father won a series of scholarships and prizes, which enabled him to send money back to his parents. He then went into research in tropical medicine, and in 1937 he traveled to East Africa as part of that research. When the war began, he made an overland journey across Africa and down the Congo River to get a ship back to England, where he volunteered for military service. He was told, however, that he was more valuable in medical research.

My father and I

With my mother

My mother was born in Dunfermline, Scotland, the third of eight children of a family doctor. The eldest was a girl with Down syndrome, who lived separately with a caregiver until she died at the age of thirteen. The family moved south to Devon when my mother was twelve. Like my father's family, hers was not well off. Nevertheless, they too managed to send my mother to Oxford. After

Oxford, she had various jobs, including that of inspector of taxes, which she did not like. She gave that up to become a secretary, which was how she met my father in the early years of the war.

I WAS born on January 8, 1942, exactly three hundred years after the death of Galileo. I estimate, however, that about two hundred thousand other babies were also born that day. I don't know whether any of them was later interested in astronomy.

I was born in Oxford, even though my parents were living in London. This was because during World War II, the Germans had an agreement that they would not bomb Oxford and Cambridge, in return for the British not bombing Heidelberg and Göttingen. It is a pity that this civilized sort of arrangement couldn't have been extended to more cities.

We lived in Highgate, in north London. My sister Mary was born eighteen months after me, and I'm told I did not welcome her arrival. All through our childhood there was a certain tension between us, fed by the narrow difference in our ages. In our adult life, however, this tension has disappeared, as we have gone different ways. She became a doctor, which pleased my father.

My sister Philippa was born when I was nearly five and better able to understand what was happening. I

Me, Philippa, and Mary

can remember looking forward to her arrival so that there would be three of us to play games. She was a very intense and perceptive child, and I always respected her judgment and opinions. My brother, Edward, was adopted much later, when I was fourteen, so he hardly entered my childhood at all. He was very different from the other three children, being completely non-academic and non-intellectual, which was probably good for us. He was a rather difficult child, but one couldn't help liking him. He died in 2004 from a cause that was never properly determined; the most likely explanation is that he was poisoned by fumes from the glue he was using for renovations in his flat.

My siblings and me at the beach

MY EARLIEST memory is of standing in the nursery of Byron House School in Highgate and crying my head off. All around me, children were playing with what seemed like wonderful toys, and I wanted to join in. But I was only two and a half, this was the first time I had been left with people I didn't know, and I was scared. I think my parents were rather surprised at my reaction, because I was their first child and they had been following child development textbooks that said that children ought to be ready to start making social relationships at two. But they took me away after that awful morning and didn't send me back to Byron House for another year and a half.

At that time, during and just after the war, Highgate was an area in which a number of scientific and academic people lived. (In another country they would have been called intellectuals, but the English have never admitted to having any intellectuals.) All these parents sent their children to Byron House School, which was a very progressive school for those times.

I remember complaining to my parents that the school wasn't teaching me anything. The educators at Byron House didn't believe in what was then the accepted way of drilling things into you. Instead, you were supposed to learn to read without realizing you were being taught. In the end, I did learn to read, but not until the fairly late age of eight. My sister Philippa was taught to read by more conventional methods and could read by the age of four. But then, she was definitely brighter than me.

We lived in a tall, narrow Victorian house, which my parents had bought very cheaply during the war, when everyone thought London was going to be bombed flat. In fact, a V-2 rocket landed a few houses away from ours. I was away with my mother and sister at the time, but my father was in the house. Fortunately, he was not hurt, and the house was not badly damaged. But for years there was a large bomb site down the road, on which I used to play with my friend Howard, who lived three doors the other way. Howard was a revelation to me because

Our street in Highgate, London

his parents weren't intellectuals like the parents of all the other children I knew. He went to the council school, not Byron House, and he knew about football and boxing, sports that my parents wouldn't have dreamed of following.

ANOTHER EARLY memory was getting my first train set. Toys were not manufactured during the war, at least not for the home market. But I had a passionate interest in model trains. My father tried making me a wooden train, but that didn't satisfy me, as I wanted

London during the Blitz

something that moved on its own. So he got a second-hand clockwork train, repaired it with a soldering iron, and gave it to me for Christmas when I was nearly three. That train didn't work very well. But my father went to America just after the war, and when he came back on the *Queen Mary* he brought my mother some nylons, which were not obtainable in Britain at that time. He brought my sister Mary a doll that closed its eyes when you laid it down. And he brought me an American train, complete with a cowcatcher and a figure-eight track. I can still remember my excitement as I opened the box.

Me with my train set

Clockwork trains, which you had to wind up, were all very well, but what I really wanted were electric trains. I used to spend hours watching a model railway club layout in Crouch End, near Highgate. I dreamed about electric trains. Finally, when both my parents were away somewhere, I took the opportunity to draw out of the Post Office bank all of the very modest amount of money that people had given me on special occasions, such as my christening. I used the money to buy an electric train set, but frustratingly enough, it didn't work very well either. I should have taken the set back and demanded that the shop or manufacturer replace it, but in those days the attitude was that it was a privilege to buy something, and it was just your bad luck if it turned out to be faulty. So I paid for the electric motor of the engine to be serviced, but it never worked very well, even then.

Later on, in my teens, I built model airplanes and boats. I was never very good with my hands, but I did this with my school friend John McClenahan, who was much better and whose father had a workshop in their house. My aim was always to build working models that I could control. I didn't care what they looked like. I think it was the same drive that led me to invent a series of very complicated games with another school friend, Roger Ferneyhough. There was a manufacturing game, complete with factories in which units of different colors were made, roads and railways on which they were carried, and a stock market. There was a war game, played on a board of four thousand squares, and even a feudal game, in which each player was a whole dynasty, with a family tree. I think these games, as well as the trains, boats, and airplanes, came from an urge to know how systems worked and how to control them. Since I began my PhD, this need has been met by my research into cosmology. If you understand how the universe operates, you control it, in a way.

2

ST. ALBANS

I N 1950 MY FATHER'S PLACE OF WORK MOVED FROM Hampstead, near Highgate, to the newly constructed National Institute for Medical Research in Mill Hill, on the northern edge of London. Rather than travel out from Highgate, it seemed more sensible for him to move the family out of London and travel into town for work. My parents therefore bought a house in the cathedral city of St. Albans, about ten miles north of Mill Hill and twenty miles north of central London. It was a large Victorian house of some elegance and character. My parents were not very well off when they bought it, and they had

Our house in St. Albans

to have quite a lot of work done on it before we could move in. Thereafter my father, like the Yorkshireman he was, refused to pay for any further repairs. Instead, he did his best to keep it up and keep it painted, but it was a big house and he was not very skilled in such matters. The house was solidly built, however, so it withstood this neglect. My parents sold it in 1985, when my father was very ill, a year before he died. I saw it recently—it didn't seem that any more work had been done on it.

The house had been designed for a family with servants, and in the pantry there was an indicator board that showed which room a bell had been rung from. Of course we didn't have servants, but my first bedroom was a little L-shaped space that must have been a maid's room. I asked for it at the suggestion of my cousin Sarah, who was slightly older than me and whom I greatly admired. She said that we could have great fun there. One of the attractions of the room was that one could climb out the window onto the roof of the bicycle shed, and thence to the ground.

Sarah was the daughter of my mother's eldest sister, Janet, who had trained as a doctor and was married to a psychoanalyst. They lived in a rather similar house in Harpenden, a village five miles farther north. They were one of the reasons we moved to St. Albans. It was a great bonus to me to be near Sarah, and I frequently went on the bus to Harpenden to see her.

St. Albans itself stood next to the remains of the ancient Roman city of Verulamium, which had been the most important Roman settlement in Britain after London. In the Middle Ages it had had the richest monastery in Britain. It was built around the shrine of Saint Alban, a Roman centurion who is said to have been the first person in Britain to be executed for his Christian faith. All that remained of the abbey was a very large and rather ugly church and the old gateway building, which was now part of St. Albans School, where I later went. St. Albans was a somewhat stodgy and conservative place compared with Highgate or Harpenden. My parents made hardly any friends there. In part this was their own fault, as they were naturally rather solitary, particularly my father. But it also reflected a different kind of population; certainly, none of the parents of my school friends in St. Albans could be described as intellectuals.

In Highgate our family had seemed fairly normal, but in St. Albans I think we were definitely regarded as eccentric. This perception was increased by the behavior of my father, who cared nothing for appearances if this allowed him to save money. His family had been very poor when he was young, and it had left a lasting impression on him. He couldn't bear to spend money on his own comfort, even when, in later years, he could afford to. He refused to put in central heating, even though he felt the cold badly. Instead he would wear several sweaters and a dressing gown on top of his normal clothes. He was, however, very generous to other people.

In the 1950s he felt we couldn't afford a new car, so he bought a prewar London taxi, and he and I built a Nissen hut as a garage. The neighbors were outraged, but they couldn't stop us. Like most boys, I was embarrassed by my parents. But it never worried them.

For holidays, my parents bought a Gypsy caravan, which they placed in a field at Osmington Mills, on the south coast of Britain near Weymouth. The caravan had

Our Gypsy caravan

been brightly and elaborately decorated by its original Gypsy owners. My father painted it green all over to make it less obvious. The caravan had a double bed for the parents and a cupboard underneath for the children, but my father converted it to bunk beds using army-surplus stretchers, while my parents slept in an army-surplus tent next door. We had our summer holidays there until 1958, when the county council finally managed to remove the caravan.

WHEN WE first came to St. Albans, I was sent to the High School for Girls, which despite its name took boys up to the age of ten. After I had been there one term, however, my father made one of his almost yearly visits to Africa, this time for a rather long period, about four months. My mother didn't feel like being left alone all that time, so she took my two sisters and me to visit her school friend Beryl, who was married to the poet Robert Graves. They lived in a village called Deya, on the Spanish island of Majorca. This was only five years after the war, and Spain's dictator, Francisco Franco, who had been an ally of Hitler and Mussolini, was still in power. (In fact, he remained in power for another two decades.) Nevertheless, my mother, who had been a member of the Young Communist League before the war, went with three young children by boat and train to Majorca. We

Me sailing on Oulton Broad, Suffolk

rented a house in Deya and had a wonderful time. I shared a tutor with Robert's son William.

This tutor was a protégé of Robert's and was more interested in writing a play for the Edinburgh festival than in teaching us. To keep us occupied, he therefore set us to read a chapter of the Bible each day and write a piece on

Our temporary home: Deya, Majorca

Me (at left) with Robert Graves's son William

it. The idea was to teach us the beauty of the English language. We got through all of Genesis and part of Exodus before I left. One of the main things I learned from this exercise was not to begin a sentence with "And." When I pointed out that most sentences in the Bible began with "And," I was told that English had changed since the time of King James. In that case, I argued, why make us read the Bible?

But it was in vain. Robert Graves at that time was very keen on the symbolism and mysticism in the Bible. So there was no one to appeal to.

We got back as the Festival of Britain was beginning. This was the Labour government's idea to try to re-create the success of the Great Exhibition of 1851, which had been organized by Prince Albert, and which was the first World's Fair in the modern sense. It provided a welcome relief from the austerity of the war and postwar years in Britain. The exhibition, held on the south bank of the Thames, opened my eyes to new forms of architecture and to new science and technology. However, the exhibition was short-lived: the Conservatives won an election that autumn and closed it down.

At the age of ten, I took the so-called eleven-plus exam. This was an intelligence test that was meant to sort out the children suited to an academic education from the majority who were sent to non-academic secondary schools. The eleven-plus system led to a number of working-class and lower-middle-class children reaching university and distinguished positions, but there was an outcry against the whole principle of a once-and-for-all selection at age eleven, mainly from middle-class parents who found their children sent to schools with working-class kids. The system was largely abandoned in the 1970s in favor of comprehensive education.

English education in the 1950s was very hierarchical. Not only were schools divided into academic and non-academic, but the academic schools were further divided into A, B, and C streams. This worked well for those in the A stream but not so well for those in the B stream and badly for those in the C stream, who got discouraged. I was put in the A stream of St. Albans School, based on the results of the eleven-plus. But after the first year, everyone who ranked below twentieth in the class was assigned to the B stream. This was a tremendous blow to their self-confidence, from which some never recovered. In my first two terms at St. Albans, I came in twenty-fourth and twenty-third, respectively, but in my third term I came in eighteenth. So I just escaped being moved down at the end of the year.

WHEN I was thirteen, my father wanted me to try for Westminster School, one of Britain's main public schools (what in the United States are called private schools). At that time, as I've mentioned, there was a sharp division in education along class lines, and my father felt that the social graces such a school would give me would be an advantage in life. My father believed that his own lack of poise and connections had led to him being passed over in his career in favor of people of less ability. He had a bit of a chip on his shoulder because he felt that other people who were not as good but who had the right background and connections had gotten ahead of him. He used to warn me against such people.

Because my parents were not well off, I would have to win a scholarship in order to attend Westminster. I was ill at the time of the scholarship examination, however, and did not take it. Instead, I remained at St. Albans School, where I got an education that was as good as, if not better than, the one I would have had at Westminster. I have never found that my lack of social graces has been a hindrance. But I think physics is a bit different from medicine. In physics it doesn't matter what school you went to or to whom you are related. It matters what you do.

I was never more than about halfway up the class. (It was a very bright class.) My classwork was very untidy, and my handwriting was the despair of my teachers. But my classmates gave me the nickname Einstein, so presumably they saw signs of something better. When I was twelve, one of my friends bet another friend a bag of

Me (at right), in my late teens

sweets that I would never amount to anything. I don't
know if this bet was ever settled, and if so, which way it
was decided.

I had six or seven close friends, most of whom I'm still
in touch with. We used to have long discussions and arguments about everything from radio-controlled models
to religion and from parapsychology to physics. One of
the things we talked about was the origin of the universe

and whether it had required a God to create it and set it going. I had heard that light from distant galaxies was shifted toward the red end of the spectrum and that this was supposed to indicate that the universe was expanding. (A shift toward the blue would have meant it was contracting.) But I was sure there must be some other reason for the red shift. An essentially unchanging and everlasting universe seemed so much more natural. Maybe light just got tired, and more red, on its way to us, I speculated. It was only after about two years of PhD research that I realized I had been wrong.

MY FATHER was engaged in research on tropical diseases, and he used to take me around his laboratory in Mill Hill. I enjoyed this, especially looking through microscopes. He also used to take me into the insect house, where he kept mosquitoes infected with tropical diseases. This worried me, because there always seemed to be a few mosquitoes flying around loose. He was very hardworking and dedicated to his research.

I was always very interested in how things operated, and I used to take them apart to see how they worked, but I was not so good at putting them back together again. My practical abilities never matched up to my theoretical inquiries. My father encouraged my interest in science, and he even coached me in mathematics until I

My father on one of his field research trips to study tropical medicine

got to a stage beyond his knowledge. With this background, and my father's job, I took it as natural that I would go into scientific research.

When I came to the last two years of school, I wanted to specialize in mathematics and physics. There was an inspiring math teacher, Mr. Tahta, and the school had also just built a new math room, which the math set had as their classroom. But my father was very much against it, because he thought there wouldn't be any jobs for mathematicians except as teachers. He would really have liked me to do medicine, but I showed no interest in biology, which seemed to me to be too descriptive and not sufficiently fundamental. It also had a rather low status at school. The brightest boys did mathematics and physics; the less bright did biology.

My father knew I wouldn't do biology, but he made me do chemistry and only a small amount of mathematics. He felt this would keep my scientific options open. I'm now a professor of mathematics, but I have not had any formal instruction in mathematics since I left St. Albans School at the age of seventeen. I have had to pick up what I know as I went along. I used to supervise undergraduates at Cambridge and keep one week ahead of them in the course.

Physics was always the most boring subject at school because it was so easy and obvious. Chemistry was much

Me (at far left) at St. Albans School

more fun because unexpected things, such as explosions, kept happening. But physics and astronomy offered the hope of understanding where we came from and why we are here. I wanted to fathom the depths of the universe. Maybe I have succeeded to a small extent, but there's still plenty I want to know.

3

OXFORD

M Y FATHER WAS VERY KEEN THAT I SHOULD GO TO
Oxford or Cambridge. He himself had gone to University College, Oxford, so he thought I should apply there, because I would have a greater chance of getting in. At that time, University College had no fellow in mathematics, which was another reason he wanted me to do chemistry: I could try for a scholarship in natural science rather than in mathematics.

The rest of the family went to India for a year, but I had to stay behind to do A-levels and university entrance exams. I stayed with the family of Dr. John Humphrey, a colleague of my father's at the National Institute for Medical Research, at their house in Mill Hill. The house had a basement that contained steam engines and other models made by John Humphrey's father, and I spent much of my time there. In the summer holidays I went to India to join the rest of the family, who were living in a house in Lucknow rented from a former chief minister of the Indian state of Uttar Pradesh who had been disgraced for corruption. My father refused to eat Indian food during his time there, so he hired an ex–British Indian Army cook and bearer to prepare and serve English food. I would have preferred something more exciting.

We went to Kashmir and rented a houseboat on the lake in Srinagar. We went in the monsoon, and the road that the Indian army had built over the mountains was washed away in some places (the normal route led across the ceasefire line to Pakistan). Our car, which we had brought from England, couldn't cope with more than three inches of water, so we had to be towed by a Sikh truck driver.

MY HEADMASTER thought I was much too young to try for Oxford, but I went up in March 1959 to do the scholarship exam with two boys from the year above me at school. I was convinced I had done badly and was very depressed when, during the practical exam,

Me coxing for the Boat Club

university lecturers came around to talk to other students but not to me. Then, a few days after I got back from Oxford, I got a telegram to say I had received a scholarship.

I was seventeen, and most of the other students in my year had done military service and were a lot older. I felt rather lonely during my first year and part of the second. In my third year, in order to make more friends, I joined the Boat Club as a coxswain. My coxing career was fairly disastrous, though. Because the river at Oxford is too narrow to race side by side, they have bumping races in which the eights line up one behind another, with each cox holding a starting line to keep his boat the right distance behind the boat ahead.

The Boat Club at rest

In my first race I let go of the starting line when the starting gun fired, but it caught in the rudder lines, with the result that our boat went off course and we were disqualified. Later I had a head-on collision with another eight, but at least in this case I can claim it was not my fault, as I had right of way over the other eight. Despite my lack of success as a cox, I did make more friends that year and was much happier.

The prevailing attitude at Oxford at that time was very anti-work. You were supposed to either be brilliant

33

The Boat Club at play

without effort or accept your limitations and get a fourth-class degree. To work hard to get a better class of degree was regarded as the mark of a "gray man," the worst epithet in the Oxford vocabulary.

The colleges at that time regarded themselves as in loco parentis (in the place of parents), which meant they were responsible for the morals of the students. The colleges were therefore all single-sex and the gates were locked at midnight, by which time all visitors—especially those of the opposite sex—were supposed to be out. After that, if you wanted to leave, you had to climb a high wall topped with spikes. My college didn't want its students getting injured, so it left a gap in the spikes, and it was quite easy to climb out. It was a different matter if you were found in bed with a member of the opposite sex, in which case you were sent down—expelled—on the spot.

The lowering of the age of majority to eighteen and the sexual revolution of the 1960s changed everything, but that was after I attended Oxford.

AT THAT time, the physics course was arranged in a way that made it particularly easy to avoid work. I did one exam before I went up, then had three years at Oxford with just the final exams at the end. I once calculated that I did about a thousand hours' work in the three years I was there, an average of an hour a day. I'm not proud of this lack of work, but at the time I shared my attitude with most of my fellow students. We affected an air of complete boredom and the feeling that nothing was worth making an effort for. One result of my illness has been to change all that. When you are faced with the possibility of an early death, it makes you realize that life is worth living and that there are lots of things you want to do.

Because of my lack of preparation, I had planned to get through the final exam by doing problems in theoretical physics and avoiding questions that required factual knowledge. I didn't sleep the night before the exam because of nervous tension, however, so I didn't do very well. I was on the borderline between first- and second-class degrees, and I had to be interviewed by the examiners to determine which I should get. In the interview they asked me about my future plans. I replied that I wanted to do research. If they gave me a first, I told them, I would go to Cambridge. If I only got a second, I would stay in Oxford. They gave me a first.

As a backup plan, in case I wasn't able to do research, I had applied to join the civil service. Because of my feelings about nuclear weapons, I didn't want anything to do with defense. I therefore listed my preference as a job at the Ministry of Works (which at that time looked after public buildings) or one as a House of Commons clerk. In the interviews it became clear that I did not really know what a House of Commons clerk did, but despite this, I

Graduation from Oxford (*above* and *right*)

passed the interviews and all that remained was a written exam. Unfortunately, I completely forgot about it and missed the exam. The civil service selection board wrote me a nice letter saying I could try again the next year and they wouldn't hold it against me. It was lucky I didn't become a civil servant. I wouldn't have managed with my disability.

IN THE long vacation following my final exam, the college offered a number of small travel grants. I thought my chances of getting one would be greater the farther I proposed to go. So I said I wanted to go to Iran. I set out with a fellow student, John Elder, who had been before and who knew the language, Farsi. We traveled by train to Istanbul, and then to Erzurum in eastern Turkey, near Mount Ararat. After that, the railway entered Soviet territory, so we had to take an Arab bus complete with chickens and sheep to Tabriz and then Tehran.

In Tehran, John and I parted company and I traveled south with another student to Isfahan, Shiraz, and Persepolis, which was the capital of the ancient Persian kings and was sacked by Alexander the Great. I then crossed the central desert to Mashhad.

On my way home, I and my traveling companion, Richard Chiin, were caught in the Bou'in-Zahra earthquake, a magnitude 7.1 quake that killed more than twelve thousand people. I must have been near the epicenter, but I was unaware of it because I was ill and in a bus that was bouncing around on the Iranian roads. Because we did not know the language, we did not learn of the disaster for several days, which we spent in Tabriz while I recovered from severe dysentery and a broken rib from being thrown against the front seat of the bus. It was not until we reached Istanbul that we learned what had happened.

I sent a postcard to my parents, who had been anxiously waiting for word for ten days. The last they had heard, I was leaving Tehran for the disaster region on the day of the quake.

4

CAMBRIDGE

I ARRIVED IN CAMBRIDGE AS A GRADUATE STUDENT IN October 1962. I had applied to work with Fred Hoyle, the most famous British astronomer of the time, and the principal defender of the steady-state theory. I say astronomer because cosmology was at that time hardly recognized as a legitimate field. That was where I wanted to do my research, inspired by having been on a summer course with Hoyle's student Jayant Narlikar. However, Hoyle had enough students already, so to my great disappointment I was assigned to Dennis Sciama, of whom I had not heard.

It was probably for the best. Hoyle was away a lot, and I wouldn't have had much of his attention. Sciama, on the other hand, was usually around and available to talk. I didn't agree with many of his ideas, particularly on Mach's principle, the idea that objects owe their inertia to the influence of all the other matter in the universe, but that stimulated me to develop my own picture.

When I began research, the two areas that seemed most exciting were cosmology and elementary particle physics. The latter was an active, rapidly changing field that attracted most of the best minds, while cosmology and general relativity were stuck where they had been in the 1930s. Richard Feynman, a Nobel Prize winner and one of the greatest physicists of the twentieth century, has given an amusing account of attending a conference on general relativity and gravitation in Warsaw in 1962. In a letter to his wife, he said, "I am not getting anything out of the meeting. I am learning nothing. Because there are no experiments, this field is not an active one, so few of the best men are doing work in it. The result is that there are hosts of dopes here (126) and it is not good for my blood pressure.... Remind me not to come to any more gravity conferences!"

OF COURSE, I wasn't aware of all this when I began my research. But I felt that the study of elementary particles at that time was too like botany. Quantum electrodynamics—the theory of light and electrons that governs chemistry and the structure of atoms—had been worked out completely in the 1940s and 1950s. Attention had now shifted to the weak and strong nuclear forces between particles in the nucleus of an atom, but similar field theories didn't seem to work to explain them. Indeed, the Cambridge school, in particular, held that there was no underlying field theory. Instead, everything would be determined by unitarity—that is, probability conservation—and certain characteristic patterns in the scattering of particles. With hindsight, it now seems amazing that it was thought this approach would work, but I remember the scorn that was poured on the first attempts at unified field theories of the weak nuclear forces, which ultimately took its place. The analytic S-matrix work is now forgotten, and I'm very glad I didn't start my research in elementary particles. None of my work from that period would have survived.

Cosmology and gravitation, on the other hand, were neglected fields that were ripe for development at that time. Unlike with elementary particles, there was a well-defined theory—the general theory of relativity—but this was thought to be impossibly difficult. People were so pleased to find any solution of the Einstein field equations that describe the theory that they didn't ask what physical significance, if any, the solution had. This was the old school of general relativity that Feynman had encountered in Warsaw. Ironically, the Warsaw conference also marked the beginning of the renaissance of general relativity, though Feynman could be forgiven for not recognizing it at the time.

A new generation entered the field, and new centers of the study of general relativity appeared. Two of these were of particular importance to me. One was located in Hamburg, Germany, under Pascual Jordan. I never visited it, but I admired the elegant papers produced there, which were such a contrast to the previous messy work on general relativity. The other center was at King's College, London, under Hermann Bondi.

Because I hadn't done much mathematics at St. Albans or in the very easy physics course at Oxford, Sciama suggested I work on astrophysics. But having been cheated out of working with Hoyle, I wasn't going to study something boring and earthbound such as Faraday rotation. I had come to Cambridge to do cosmology, and cosmology I was determined to do. So I read old textbooks on general relativity and traveled up to lectures at King's College, London, each week, with three other students of Sciama's. I followed the words and equations, but I didn't really get a feel for the subject.

SCIAMA INTRODUCED me to the so-called Wheeler-Feynman electrodynamics. This theory said that electricity and magnetism were time symmetric. However, when one switched on a lamp, it was the influence of all the other matter in the universe that caused light waves to travel outward from the lamp, rather than come in from infinity and end on the lamp. For Wheeler-Feynman electrodynamics to work, it was necessary that all the light traveling out from the lamp should be absorbed by other matter in the universe. This would happen in a steady-state universe, in which the density of matter would remain constant, but not in a Big Bang universe, where the density would go down as the universe expanded. It was claimed that this was another proof, if proof were still needed, that we live in a steady-state universe.

This was supposed to explain the arrow of time, the reason disorder increases and why we remember the past but not the future. There was a conference on Wheeler-Feynman electrodynamics and the arrow of time at Cornell University in 1963. Feynman was so disgusted by the nonsense put forth about the arrow of time that he refused to let his name appear in the proceedings. He was referred to only as Mr. X, but everyone knew who that was.

I found that Hoyle and Narlikar had already worked out Wheeler-Feynman electrodynamics in expanding universes and then gone on to formulate a time-symmetric new theory of gravity. Hoyle unveiled the theory at a meeting of the Royal Society in 1964. I was at the lecture, and in the question period I said that the influence of all the matter in a steady-state universe would make his masses infinite. Hoyle asked why I said that, and I replied that I had calculated it. Everyone thought I meant that I had done it in my head during the lecture, but in fact I had been sharing an office with Narlikar and had seen a draft of the paper ahead of time, which had allowed me to do the calculations before the meeting.

Hoyle was furious. He was trying to set up his own institute, and threatened to join the brain drain to America if he didn't get the money. He thought I had been put up to it to sabotage his plans. However, he got his institute, and later gave me a job, so he apparently didn't harbor a grudge against me.

IN MY last year at Oxford I noticed that I was getting increasingly clumsy. I went to the doctor after falling down some stairs, but all he said was "Lay off the beer."

I became even more clumsy after moving to Cambridge. At Christmas, when I went skating on the lake at St. Albans, I fell over and couldn't get up. My mother noticed these problems and took me to the family doctor. He referred me to a specialist, and shortly after my twenty-first birthday I went into the hospital for testing. I was in for two weeks, during which I had a wide variety of tests. They took a muscle sample from my arm, stuck electrodes into me, and then injected some radio-opaque fluid into my spine and with X-rays watched it go up and down as they tilted the bed. After all that, they didn't tell me what I had, except that it was not multiple sclerosis and that I was an atypical case. I gathered, however, that they expected it to continue to get worse and that there was nothing they could do except give me vitamins, though I could see they didn't expect them to have much effect. I didn't ask for more details at that time, because they obviously had nothing good to tell me.

The realization that I had an incurable disease that was likely to kill me in a few years was a bit of a shock. How could something like this happen to me? However, while I was in the hospital, I had seen a boy I vaguely knew die of leukemia in the bed opposite me, and it had not been a pretty sight. Clearly there were people who were worse off than me—at least my condition didn't make me feel sick. Whenever I feel inclined to be sorry for myself, I remember that boy.

NOT KNOWING what was going to happen to me or how rapidly the disease would progress, I was at a loose end. The doctors told me to go back to Cambridge and carry on with the research I had just started in general relativity and cosmology. But I was not making progress because I didn't have much mathematical background—and anyway, it was hard to focus when I might not live long enough to finish my PhD. I felt somewhat of a tragic character.

I took to listening to Wagner, but reports in magazine articles that I also drank heavily at that time are an exaggeration. Once one article said it, other articles copied it because it made a good story, and eventually everyone believed that anything that had appeared in print so many times must be true.

My dreams at that time, however, were rather disturbed. Before my condition was diagnosed, I had been very bored with life. There had not seemed to be anything worth doing. But shortly after I came out of the hospital, I dreamed that I was going to be executed. I suddenly realized that there were a lot of worthwhile things I could do if I was reprieved. Another dream I had several times was that I would sacrifice my life to save others. After all, if I was going to die anyway, I might as well do some good.

BUT I didn't die. In fact, although there was a cloud hanging over my future, I found to my surprise that I was enjoying life. What really made the difference was that I got engaged to a girl called Jane Wilde, whom I had met about the time I was diagnosed with ALS. This gave me something to live for.

Punting on the Cam with Jane

If we were to get married, I had to get a job, and to get a job I had to finish my PhD. I therefore started working for the first time in my life. To my surprise, I found I liked it. Maybe it is not fair to call it work, though. Someone once said that scientists and prostitutes get paid for doing what they enjoy.

To support myself during my studies, I applied for a research fellowship at Gonville and Caius College, a college within the University of Cambridge. Because my increasing clumsiness made it difficult to write or type, I was hoping that Jane would type my application. But when she came to visit me in Cambridge, she had her arm in plaster, having broken it. I must admit that I was less sympathetic than I should have been. It was her left arm, however, so she was able to write out the application as I dictated it, and I got someone else to type it.

In my application I had to give the names of two people who could provide references about my work. My supervisor suggested I should ask Hermann Bondi to be one of them. Bondi was then a professor of mathematics at King's College, London, and an expert on general relativity. I had met him a couple of times, and he had submitted one of my papers for publication in the journal *Proceedings of the Royal Society.* After a lecture he gave in Cambridge, I asked him about providing a reference, and he looked at me in a vague way and said yes, he would. Obviously he didn't remember me, for when the college wrote to him for a reference, he replied that he had not heard of me. Nowadays there are so many people applying for college research fellowships that if one of the candidate's referees says he does not know him, that is the end of his chances. But those were quieter times. The college wrote to tell me of my referee's embarrassing reply, and my supervisor got on to Bondi and refreshed his memory. Bondi then wrote me a reference that was probably far better than I deserved. I got a research fellowship and have been a fellow of Caius College ever since.

The fellowship meant Jane and I could get married,

which we did in July 1965. We spent a week's honeymoon in Suffolk, which was all I could afford. We then went to a summer school in general relativity at Cornell University.

That was a mistake. We stayed in a dormitory that was full of couples with noisy small children, and it put quite a strain on our marriage. In other respects, however, the summer school was very useful for me because I met many of the leading people in the field.

When we were first married, Jane was still an undergraduate at Westfield College in London. So she had to go up to London from Cambridge during the week to

My wedding to Jane

complete her degree. The disease was causing increasing muscle weakness, which meant that it was becoming harder to walk, and so we had to find a centrally located place where I could manage on my own. I asked the college for help but was told by the bursar that it was not college policy to help fellows with housing. We therefore put our name down to rent one of a group of new flats that was being built in the marketplace, a convenient location. (Years later, I discovered that those flats were actually owned by the college, but they didn't tell me that.) When we returned to Cambridge from the summer in America, however, we found that the flats were not ready.

As a great concession, the bursar offered us a room in a hostel for graduate students. He said, "We normally charge twelve shillings and sixpence a night for this room. However, as there will be two of you in the room, we will charge twenty-five shillings." We stayed there only three nights before we found a small house about one hundred yards from my university department. It belonged to another college, which had let it to one of its fellows. He had recently moved out to a house in the suburbs, and he sublet the house to us for the remaining three months on his lease.

During those three months, we found another house in the same road standing empty. A neighbor summoned the owner from Dorset and told her it was a scandal that her house should be vacant when young people were looking for accommodation, so she rented the house to us. After we had lived there for a few years, we wanted to buy it and fix it up, so we asked my college for a mortgage. The college did a survey and decided it was not a good risk, so in the end we got a mortgage elsewhere, and my parents gave us the money to renovate.

THE SITUATION in Caius College was at that time reminiscent of something out of the novels of C. P. Snow. There had been a bitter division in the fellowship ever since the so-called Peasants' Revolt, in which a number of the more junior fellows had banded together to vote senior fellows out of office. There were two camps: on one side was the party of the master and bursar, and on the other side was a more progressive party that wanted to spend more of the college's considerable wealth on academic purposes. The progressive party took advantage of a meeting of the college council at which the master and bursar were absent to elect six research fellows, including me.

At my first college meeting there were elections to the college council. The other new research fellows had been briefed on whom to vote for, but I was a complete innocent and voted for candidates of both parties. The progressive party won a majority on the college council, and Master Sir Nevill Mott (who later won a Nobel Prize for work in condensed-matter physics) resigned in anger.

However, the next master, Joseph Needham (author of a multivolume history of science in China), healed the wound, and the college has been relatively peaceful ever since.

OUR FIRST child, Robert, was born after we had been married about two years. Shortly after his birth we took him to a scientific meeting in Seattle. That again was a mistake. I was not able to help much with the baby because of my increasing disability, and Jane had to cope

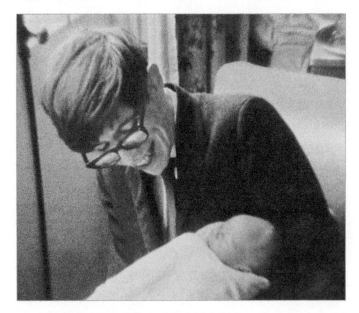

With my first child, Robert

Jane and Robert

largely on her own and got very tired. Her tiredness was compounded by further traveling we did in the United States after Seattle. Robert now lives in Seattle with his wife, Katrina, and their children, George and Rose, so obviously the experience didn't scar him.

Our second child, Lucy, was born about three years later in an old workhouse that was then being used as a maternity hospital. During the pregnancy we had to move out to a thatched cottage owned by friends while our own house was being extended. We moved back only a few days before the birth.

5

GRAVITATIONAL WAVES

I N 1969 JOSEPH WEBER REPORTED THE OBSERVATION
of bursts of gravitational waves, using detectors that
consisted of two aluminum bars suspended in a vacuum.
When a gravitational wave came along, it would stretch
things in one direction (perpendicular to the direction
of travel of the wave) and compress things in the other
direction (perpendicular to the wave). This would make
the bars oscillate at their resonant frequency—1,660 cycles
per second—and these oscillations would be detected by
crystals strapped to the bars. I visited Weber near Prince-
ton in early 1970 and inspected his equipment. With my

untrained eye I could see nothing wrong, but the results that Weber was claiming were truly remarkable. The only possible sources of bursts of gravitational waves powerful enough to excite Weber's bars would be the collapse of a massive star to form a black hole, or the collision and merger of two black holes. These sources would have to be nearby—within our galaxy. Previous estimates of such events had been about one per century, but Weber was claiming to see one or two bursts per day. This would have meant the galaxy was losing mass at a rate that could not have been continuous over the lifetime of the galaxy—or there would be no galaxy left now.

When I returned to England, I decided that Weber's amazing claims needed independent verification. I wrote a paper with my student Gary Gibbons on the theory of the detection of bursts of gravitational waves, in which we suggested a more sensitive detector design. When it seemed that no one was about to build such a detector, Gary and I took the audacious step, for theorists, of applying to the Science Research Council for a grant to build two detectors. (One needs to observe coincidences between at least two detectors because of spurious signals from noise and earth vibrations.) Gary scoured war-surplus dumps for decompression chambers to use as vacuums and I looked for suitable sites.

Eventually we had a meeting with other groups interested in verifying Weber's claims at the Science Research

Council on the thirteenth floor of a tower block in London. (The Science Research Council couldn't admit to superstition. They got it cheap.) As there were other groups pursuing the project, Gary and I withdrew our application. That was a narrow escape! My increasing disability would have made me hopeless as an experimenter. And it is very difficult to make a mark for oneself in an experimental subject. One is often only part of a large team, doing an experiment that takes years. On the other hand, a theorist can have an idea in a single afternoon, or, in my case, while getting into bed, and write a paper on one's own or with one or two colleagues to make one's name.

Gravitational wave detectors have become much more sensitive since the 1970s. The current detectors employ laser ranging to compare the lengths of two arms at right angles. The U.S. has two of these LIGO detectors. Although they are ten million times more sensitive than Weber's, they have not so far made a reliable detection of gravitational waves. I'm very glad I remained a theorist.

6

THE BIG BANG

THE BIG QUESTION IN COSMOLOGY IN THE EARLY 1960s was whether the universe had a beginning. Many scientists were instinctively opposed to the idea, and thus to the Big Bang theory, because they felt that a point of creation would be a place where science broke down. One would have to appeal to religion and the hand of God to determine how the universe started off.

Two alternative scenarios were therefore put forward. One was the steady-state theory, in which, as the universe expanded, new matter was continually created to keep the density constant on average. The steady-state theory never had a very strong theoretical basis, because it required a negative energy field to create the matter. This would have made it unstable and prone to runaway production of matter and negative energy. But it had the great merit of making definite predictions that could be tested by observations.

By 1963 the steady-state theory was already in trouble. Martin Ryle's radio astronomy group at the Cavendish Laboratory did a survey of faint radio sources and found that the sources were distributed fairly uniformly across the sky. This indicated that they were probably outside our galaxy, because otherwise they would be concentrated along the Milky Way. But the graph of the number of sources against source strength did not agree with the prediction of the steady-state theory. There were too many faint sources, indicating that the density of sources had been higher in the distant past.

Hoyle and his supporters put forward increasingly contrived explanations of the observations, but the final nail in the coffin of the steady-state theory came in 1965 with the discovery of a faint background of microwave radiation. (This is like the microwaves in a microwave oven but at a much lower temperature, only 2.7 kelvin, a small amount above absolute zero.) The radiation could not be accounted for in the steady-state theory, though Hoyle and Narlikar tried desperately. It was just as well I hadn't been a student of Hoyle's, because I would have had to defend the steady-state theory.

The microwave background indicated that the universe had had a hot, dense stage in the past. But it didn't prove that this stage was the beginning of the universe. One might imagine that the universe had had a previous contracting phase and that it had bounced from contraction to expansion, at a high but finite density. Whether that was in fact the case was clearly a fundamental question, and it was just what I needed to complete my PhD thesis.

Gravity pulls matter together, but rotation throws it apart. So my first question was whether rotation could cause the universe to bounce. Together with George Ellis, I was able to show that the answer was no if the universe was spatially homogeneous—that is, if it was the same at each point of space. However, two Russians, Evgeny Lifshitz and Isaak Khalatnikov, claimed to have proved that a general contraction without exact symmetry would always lead to a bounce, with the density remaining finite. This result was very convenient for Marxist-Leninist dialectical materialism because it avoided awkward questions about the creation of the universe. It therefore became an article of faith for Soviet scientists.

Lifshitz and Khalatnikov were members of the old school in general relativity—that is, they wrote down a massive system of equations and tried to guess a solution.

INTRODUCTION

The idea that the universe is expanding is of recent
origin. All the early cosmologies were essentially
stationary and even Einstein whose theory of relativity is
the basis for almost all modern developments in cosmology,
found it natural to suggest a static model of the universe.
However there is a very grave difficulty associated with a
static model such as Einstein's which is supposed to have
existed for an infinite time. For, if the stars had been r

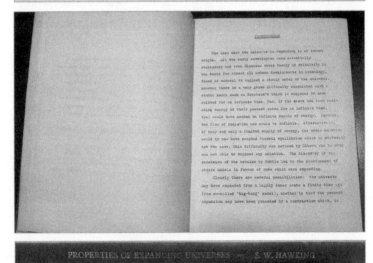

PROPERTIES OF EXPANDING UNIVERSES ~ S. W. HAWKING

My dissertation, finished at long last

But it wasn't clear that the solution they found was the most general one. Roger Penrose introduced a new approach, which didn't require solving Einstein's field equations explicitly, just certain general properties, such as that energy is positive and gravity is attractive. Penrose gave a seminar on the subject at King's College, London, in January 1965. I wasn't at the seminar, but I heard about it from Brandon Carter, with whom I shared an office in Cambridge's new Department of Applied Mathematics and Theoretical Physics (DAMTP) premises in Silver Street.

At first I couldn't understand what the point was. Penrose had shown that once a dying star contracted to a certain radius, there would inevitably be a singularity, a point where space and time came to an end. Surely, I thought, we already knew that nothing could prevent a massive cold star from collapsing under its own gravity until it reached a singularity of infinite density. But in fact the equations had been solved only for the collapse of a *perfectly spherical* star, and of course a real star won't be exactly spherical. If Lifshitz and Khalatnikov were right, the departures from spherical symmetry would grow as the star collapsed, and would cause different parts of the star to miss each other, thus avoiding a singularity of infinite density. But Penrose showed they were wrong: small departures from spherical symmetry will not prevent a singularity.

I realized that similar arguments could be applied to

the expansion of the universe. In this case, I could prove there were singularities where space-time had a beginning. So again Lifshitz and Khalatnikov were wrong. General relativity predicted that the universe should have a beginning, a result that did not pass unnoticed by the Church.

The original singularity theorems of both Penrose and myself required the assumption that the universe had a Cauchy surface, that is, a surface that intersects every particle path once and only once. It was therefore possible that our first singularity theorems simply proved that the universe didn't have a Cauchy surface. While interesting, this didn't compare in importance with time having a beginning or end. I therefore set about proving singularity theorems that didn't require the assumption of a Cauchy surface.

In the next five years, Roger Penrose, Bob Geroch, and I developed the theory of causal structure in general relativity. It was a wonderful feeling, having a whole field virtually to ourselves. How unlike particle physics, where people were falling over themselves to latch on to the latest idea. They still are.

I wrote up some of this in an essay that won an Adams Prize at Cambridge in 1966. This was the basis for the book *The Large Scale Structure of Space-Time,* which I wrote with George Ellis and which was published by Cambridge University Press in 1973. The book is still in print because it is virtually the last word on the causal structure of space-time: that is, which pole of space-time can affect events at other points. I would caution the general reader against attempting to consult it. It is highly technical and was written at a time when I was trying to be as rigorous as a pure mathematician. Nowadays I'm concerned to be right rather than righteous. Anyway, it is almost impossible to be rigorous in quantum physics, because the whole field is on very shaky mathematical ground.

7

BLACK HOLES

THE IDEA BEHIND BLACK HOLES GOES BACK MORE than two hundred years. In 1783 a Cambridge don, John Michell, published a paper in *Philosophical Transactions of the Royal Society of London* about what he called "dark stars." He pointed out that a star that was sufficiently massive and compact would have such a strong gravitational field that light could not escape. Any light emitted from the surface of the star would be dragged back by the star's gravitational attraction before it could get very far.

Michell suggested that there might be a large number

of stars like this. Although we would not be able to see them, because the light from them would not reach us, we would still feel their gravitational attraction. Such objects are what we now call black holes, because that is what they are: black voids in space. A similar suggestion was made a few years later by a French scientist, the Marquis de Laplace, apparently independently of Michell. Interestingly enough, Laplace included it in only the first and second editions of his book *The System of the World* and left it out of later editions. Perhaps he decided that it was a crazy idea.

Both Michell and Laplace thought of light as consisting of particles, rather like cannonballs, that could be slowed down by gravity and made to fall back on the star. This was not consistent with the Michelson-Morley experiment, carried out in 1887, which showed that light always travels at the same speed. A consistent theory of how gravity affects light did not come until 1915, when Einstein formulated general relativity. Using general relativity, Robert Oppenheimer and his students George Volkoff and Hartland Snyder showed in 1939 that a star that had exhausted its nuclear fuel could not support itself against gravity if its mass was greater than a certain limit, about the order of the mass of the Sun. Burnt-out stars above this mass would collapse in on themselves and form black holes containing singularities of infinite density. Although they were a prediction of his theory,

Einstein never accepted black holes or that matter could be compressed to infinite density.

Then the war intervened and diverted Oppenheimer to work on the atomic bomb. After the war, people were more interested in atomic and nuclear physics and neglected gravitational collapse and black holes for more than twenty years.

INTEREST IN gravitational collapse was reawakened in the early 1960s with the discovery of quasars, very distant objects that are very compact and powerful optical and radio sources. Matter falling into a black hole was the only plausible mechanism that could explain the production of so much energy in so small a region of space. Oppenheimer's work was rediscovered and people began to work on the theory of black holes.

In 1967 Werner Israel produced an important result. He showed that unless the remnant from a non-rotating collapsing star was exactly spherical, the singularity it contained would be naked—that is, it would be visible to outside observers. This would have meant the breakdown of general relativity at the singularity of a collapsing star, destroying our ability to predict the future of the rest of the universe.

At first, most people, including Israel himself, thought this implied that because real stars aren't exactly spherical, their collapse would give rise to naked singularities and a breakdown of predictability. However, a different interpretation was put forward by Roger Penrose and John Wheeler: that the remnant from the gravitational collapse of a non-rotating star would rapidly settle down to a spherical state. They suggested that there is cosmic censorship: nature is a prude and hides singularities in black holes, where they can't be seen.

I used to have a bumper sticker that read BLACK HOLES ARE OUT OF SIGHT on the door of my office in DAMTP. This so irritated the head of the department that he engineered my election to the Lucasian Professorship, moved me to a better office on the strength of it, and personally tore the offending notice off the door of the old office.

MY WORK on black holes began with a eureka moment in 1970, a few days after the birth of my daughter, Lucy. While getting into bed, I realized that I could apply to black holes the causal structure theory I had developed for singularity theorems. In particular, the area of the horizon, the boundary of the black hole, would always increase. When two black holes collide and merge, the area of the final black hole is greater than the sum of the areas of the original holes. This, and other properties that Jim Bardeen, Brandon Carter, and I discovered, suggested that the area was like the entropy of a black hole.

This would be a measure of how many states a black hole could have on the inside for the same appearance on the outside. But the area couldn't actually be the entropy, because if black holes had entropy, they would also have a temperature and would glow like a hot body. As everyone thought, black holes were completely black and didn't emit light or anything else.

There was an exciting period culminating in the Les Houches summer school in 1972 in which we solved most of the major problems in black hole theory. In particular, David Robinson and I proved the no-hair theorem, which said that a black hole would settle down to a state characterized by only two numbers, the mass and the rotation. This again suggested that black holes had entropy, because many different stars could collapse to produce a black hole of the same mass and rotation.

All this theory was developed before there was any observational evidence for black holes, which shows that Feynman was wrong when he said an active research field has to be experimentally driven. The one problem that was never solved was to prove the cosmic censorship hypothesis, though a number of attempts to disprove it failed. It is fundamental to all work on black holes, so I have a strong vested interest in its being true. I therefore have a bet with Kip Thorne and John Preskill on the outcome of this problem. It is difficult for me to win this bet, but quite possible for me to lose if anyone finds a counterexample with a naked singularity. In fact, I lost an earlier version of the bet, by not being careful enough about the wording. Thorne and Preskill were not amused by the T-shirt I offered in settlement.

Cosmology humor, part one:
I had this printed on a T-shirt to settle a bet

WE WERE so successful with the classical general theory of relativity that I was at a bit of a loose end in 1973, after the publication of *The Large Scale Structure of Space-Time*. My work with Penrose had shown that general relativity would break down at singularities. So the obvious next step would be to combine general relativity, the theory of the very large, with quantum theory, the theory of the very small. I had no background in quantum theory, and the singularity problem seemed too difficult for a frontal assault at that time. So as a warm-up exercise, I considered how particles and fields governed by quantum theory would behave near a black hole. In particular, I wondered, can one have atoms in which the nucleus is a tiny primordial black hole, formed in the early universe?

To answer this, I studied how quantum fields would scatter off a black hole. I was expecting that part of an incident wave would be absorbed and the remainder scattered. But to my great surprise, I found that there seemed to be emission from the black hole. At first I thought this must be a mistake in my calculation. What finally persuaded me that it was real was that the emission was exactly what was required to identify the area of the horizon with the entropy of a black hole. It is summed up in this simple formula:

$$S = \frac{Ac^3}{4\hbar G}$$

where S is the entropy and A is the area of horizon. This expression contains the three fundamental constants of nature: c, the speed of light; G, Newton's constant of gravitation; and $\hbar$ *bar,* Planck's constant. It reveals that there is a deep and previously unsuspected relationship between gravity and thermodynamics, the science of heat.

The radiation from a black hole will carry away energy, so the black hole will lose mass and shrink. Eventually, it seems, the black hole will evaporate completely and disappear. This raised a problem that struck at the heart of physics. My calculation suggested that the radiation was exactly thermal and random, as it has to be if the area of the horizon is to be the entropy of the black hole. So how could the radiation left over carry all the information about what made the black hole? Yet if information is lost, this is incompatible with quantum mechanics.

This paradox had been argued for thirty years, without much progress, until I found what I think is its resolution. Information is not lost, but it is not returned in a useful way. It is like burning an encyclopedia: the information contained in the encyclo-

Whereas Stephen Hawking and Kip Thorne firmly believe that information swallowed by a black hole is forever hidden from the outside universe, and can never be revealed even as the black hole evaporates and completely disappears,

And whereas John Preskill firmly believes that a mechanism for the information to be released by the evaporating black hole must and will be found in the correct theory of quantum gravity,

Therefore Preskill offers, and Hawking/Thorne accept, a wager that:

When an initial pure quantum state undergoes gravitational collapse to form a black hole, the final state at the end of black hole evaporation will always be a pure quantum state.

The loser(s) will reward the winner(s) with an encyclopedia of the winner's choice, from which information can be recovered at will.

Stephen W. Hawking & Kip S. Thorne   John P. Preskill

Pasadena, California, 6 February 1997

Cosmology humor, part two: a bet with John Preskill

pedia is not technically lost if one keeps all the smoke and ashes, but it is very hard to read. In fact, Kip Thorne and I had a bet with John Preskill on the information paradox. When John won the bet, I gave him a baseball encyclopedia, but maybe I should have just given him the ashes.

8

CALTECH

I N 1974 I WAS ELECTED A FELLOW OF THE ROYAL Society. The election came as a surprise to members of my department because I was young and only a lowly research assistant. But within three years I had been promoted to professor.

Jane became depressed after my election, feeling I had achieved my goals and that it was going to be downhill after that. Her depression was lifted somewhat when my friend Kip Thorne invited us and a number of others working in general relativity to the California Institute of Technology (Caltech).

Our house in Pasadena

For the past four years, I had been using a manual wheelchair as well as a blue electric three-wheeled car, which went at a slow cycling speed, and in which I sometimes illegally carried passengers. When we went to California, we stayed in a Caltech-owned colonial-style house near the campus, and there I used an electric wheelchair for the first time. It gave me a considerable degree of independence, especially as in the United States buildings and sidewalks are much more accessible for the disabled than they are in Britain. I also had one of my research students live with us. He helped me with getting up and going to bed and some meals, in return for accommodation and a lot of my academic attention.

Jane, Lucy, Robert, and me at home in Pasadena (*above* and *right*)

Our two children at that time, Robert and Lucy, loved California. The school they attended there was afraid its students would be kidnapped, so one couldn't just collect one's child from the school gate in the normal way. Instead one had to drive around the block and come to the gate one by one. The child in question would then be summoned over a bullhorn. I'd never encountered anything like this before.

The house was equipped with a color television set. In England, we'd had only a black-and-white set that hardly worked. So we watched a lot of television, particularly

British series such as *Upstairs, Downstairs* and *The Ascent of Man*. We had just watched the episode of *The Ascent of Man* in which Galileo is tried by the Vatican and condemned to house arrest for the rest of his life when I heard that I had been awarded the Pius XI Medal by the Pontifical Academy of Sciences. At first I felt like indignantly refusing it, but then I had to admit that the Vatican had ultimately changed its mind about Galileo. So I flew to England to meet up with my parents, who then accompanied me to Rome. While visiting the Vatican, I made a point of demanding to be shown the account of the trial of Galileo in the Vatican library.

At the award ceremony, Pope Paul VI got down from

his throne and knelt by my side. After the ceremony I met Paul Dirac, one of the founders of quantum theory, to whom I had not talked while he was a professor at Cambridge because I had not at that time been interested in matters quantum. He told me he had originally proposed another candidate for the medal but in the end had decided I was better and had told the academy to award it to me.

THE TWO principal stars of the Caltech physics department at that time were the Nobel Prize winners Richard Feynman and Murray Gell-Mann, and there was great rivalry between them. At the first of Gell-Mann's weekly seminars, he said, "I'm just going to repeat some talks I gave last year," whereupon Feynman got up and walked out. Gell-Mann then said, "Now that he's gone, I can tell you what I really wanted to talk about."

This was an exciting time in particle physics. New "charmed" particles had just been discovered at Stanford, and the discovery helped confirm Gell-Mann's theory that protons and neutrons were made of three more fundamental particles called quarks.

While at Caltech, I bet Kip Thorne that the binary star system Cygnus X-1 did not contain a black hole. Cygnus X-1 is an X-ray source in which a normal star is losing its outer envelope to an unseen compact companion. As matter falls toward the companion, it develops a spiral motion and gets very hot, emitting X-rays. I was hoping to lose this bet, as I obviously had made a big intellectual investment in black holes. But if they were shown not to exist, at least I would have had the consolation of winning a four-year subscription to *Private Eye* magazine. On the other hand, if Kip won, he would receive one year of *Penthouse* magazine. In the years following the bet, the evidence for black holes became so strong that I conceded and gave Kip a subscription to *Penthouse,* much to the displeasure of his wife.

WHILE IN California, I worked with a research student at Caltech, Don Page. Don had been born and brought up in a village in Alaska where his parents were schoolteachers and the three of them were the only non-Inuits. He was an evangelical Christian, and he did his best to convert me when he later came to live with us in Cambridge. He used to read me Bible stories at breakfast, but I told him I knew the Bible well from my time in Majorca, and because my father used to read the Bible to me. (My father was not a believer but thought the King James Bible was culturally important.)

Don and I worked on whether it might be possible to observe the emission from black holes that I had predicted. The temperature of the radiation from a black hole of the mass of the Sun would be only about a millionth of a kelvin, barely above absolute zero, so it would be swamped by the cosmic background of microwaves, which has a temperature of 2.7 kelvin. However, there might be much smaller black holes left over from the Big Bang. A primordial black hole with the mass of a mountain would be emitting gamma rays and would now be ending its lifetime, having radiated away most of its original mass. We looked for evidence of such emissions in the background of gamma rays but found no sign. We were able to place an upper limit on the number density of black holes of this mass, which indicates that we are not likely to be close enough to one to detect it.

9

MARRIAGE

WHEN WE RETURNED FROM CALTECH IN 1975, WE knew that the stairs in our house would now be too difficult for me. The college by then appreciated me rather more, so it let us have a ground-floor apartment in a large Victorian house it owned. (The house has now been demolished and replaced by a student accommodation block bearing my name.) The apartment was in gardens maintained by the college gardeners, which was nice for the children.

I initially felt rather low on returning to England. Everything seemed so parochial and restricted there compared to the can-do attitude in America. At the time, the landscape was littered with dead trees killed by Dutch elm disease and the country was beset by strikes. However, my mood lifted as I saw success in my work and was elected, in 1979, to the Lucasian Professorship of Mathematics, a post once held by Sir Isaac Newton and Paul Dirac.

Our third child, Tim, was also born in 1979 after a trip to Corsica, where I was lecturing at a summer school. Thereafter Jane became more depressed. She was worried

With my family after the christening of our third child, Tim

I was going to die soon and wanted someone who would give her and the children support and marry her when I was gone. She found Jonathan Jones, a musician and organist at the local church, and gave him a room in our apartment. I would have objected, but I too was expecting an early death and felt I needed someone to support the children after I was gone.

I continued to get worse, and one of the symptoms of my progressing illness was prolonged choking fits. In 1985, on a trip to CERN (European Organization for Nuclear Research) in Switzerland, I caught pneumonia. I was rushed to the cantonal hospital and put on a ventilator. The doctors at the hospital thought I was so far gone that they offered to turn off the ventilator and end my life, but Jane refused and had me flown back by air ambulance to Addenbrooke's Hospital in Cambridge. The doctors there tried hard to get me back to how I had been before, but in the end they had to perform a tracheotomy.

Before my operation my speech had been getting more slurred, so only people who knew me well could understand me. But at least I could communicate. I wrote scientific papers by dictating to a secretary, and I gave seminars through an interpreter who repeated my words more clearly. However, the tracheotomy removed my ability to speak altogether. For a time, the only way I could communicate was to spell out words letter by letter by raising my eyebrows when someone pointed to the right letter on a spelling card. It is pretty difficult to carry on a conversation like that, let alone write a scientific paper. However, a computer expert in California named Walt Woltosz heard of my plight and sent me a computer program that he had written, called Equalizer. This allowed me to select words from a series of menus on the screen by pressing a switch in my hand. I now use another of his programs, called Words Plus, which I control by a small sensor on my glasses that responds to my cheek movement. When I have built up what I want to say, I can send it to a speech synthesizer.

At first I just ran the Equalizer program on a desktop computer. Then David Mason, of Cambridge Adaptive Communication, fitted a small personal computer and a speech synthesizer to my wheelchair. My computers are now supplied by Intel. This system allows me to communicate much better than I could before, and I can manage up to three words a minute. I can either speak what I have written or save it on disk. I can then print it out or call it back and speak it sentence by sentence. Using this system, I have written seven books and a number of scientific papers. I have also given a number of scientific and popular talks. They have been well received, which I think is due in large part to the quality of the speech synthesizer, made by Speech Plus.

One's voice is very important. If you have a slurred voice, people are likely to treat you as mentally deficient.

This synthesizer was by far the best I had heard because it varies the intonation and didn't speak like one of the Daleks from *Doctor Who*. Speech Plus has since gone into liquidation and its speech synthesizer program has been lost. I now have the last three remaining synthesizers. They are bulky, use a lot of power, and contain chips that are obsolete and can't be replaced. Nevertheless, by now I identify with the voice and it has become my trademark, so I won't change it for a more natural-sounding voice unless all three synthesizers break.

When I came out of the hospital I needed full-time nursing care. At first I felt my scientific career was over and all that would be left to me would be to stay at home and watch television. But I soon found I could carry on my scientific work and write mathematical equations using a program called Latex, which allows one to write mathematical symbols in ordinary characters, such as *$/pi$* for π.

HOWEVER, I became more and more unhappy about the increasingly close relationship between Jane and Jonathan. In the end I could stand the situation no longer, and in 1990 I moved out to a flat with one of my nurses, Elaine Mason.

We found the flat rather small for us and Elaine's two sons, who were with us for part of the week, so we decided to move. A bad storm in 1987 had torn off the roof of Newnham College, the sole women-only undergraduate college. (The men-only colleges had all by this time admitted women. My college, Caius, which had a number of conservative fellows, was one of the last, and it was finally persuaded by the students' exam results that it wouldn't get good men applying unless it admitted women as well.) Because Newnham was a poor college, it had had to sell four plots of land to pay for the roof repair after the storm. We bought one of the plots and built a wheelchair-friendly house.

Elaine and I got married in 1995. Nine months later Jane married Jonathan Jones.

My wedding to Elaine

My marriage to Elaine was passionate and tempestuous. We had our ups and downs, but Elaine's being a nurse saved my life on several occasions. After the tracheotomy, I had a plastic tube in my trachea, which prevented food and saliva from getting into my lungs and was retained by an inflated cuff. Over the years the pressure in the cuff damaged my trachea and made me cough and choke. I was coughing on a flight back from Crete, where I had been at a conference, when David Howard, a surgeon who happened to be on the same plane, approached Elaine and said he could help me. He suggested a laryngectomy, which would completely separate my windpipe from my throat and remove the need for a tube with a cuff. The doctors at Addenbrooke's Hospital in Cambridge said it was too risky, but Elaine insisted, and David Howard carried out the operation in a London hospital. That operation saved my life: another two weeks and the cuff would have worn a hole between my windpipe and my throat, filling my lungs with blood.

A few years later I had another health crisis because my oxygen levels were falling dangerously low in deep sleep. I was rushed to the hospital, where I remained for four months. I was eventually discharged with a ventilator, which I used at night. My doctor told Elaine that I was coming home to die. (I have since changed my doctor.) Two years ago I began using the ventilator twenty-four hours a day. I find it gives me energy.

With Elaine in Aspen, Colorado (*above* and *right*)

A year after that I was recruited to help the university's fund-raising campaign for its eight-hundredth anniversary. I was sent to San Francisco, where I gave five lectures in six days and got very tired. One morning I passed out when I was taken off the ventilator. The nurse on duty thought I was okay, but I would have died had not another caregiver summoned Elaine, who resuscitated me. All these crises took their emotional toll on Elaine. We got divorced in 2007, and since the divorce I have lived alone with a housekeeper.

10

A BRIEF HISTORY OF TIME

FIRST HAD THE IDEA OF WRITING A POPULAR BOOK about the universe in 1982. My intention was partly to earn money to pay my daughter's school fees. (In fact, by the time the book actually appeared, she was in her last year of school.) But the main reason for writing it was that I wanted to explain how far I felt we had come in our understanding of the universe: how we might be near finding a complete theory that would describe the universe and everything in it.

If I was going to spend the time and effort to write a book, I wanted it to get to as many people as possible. My previous technical books had been published by Cambridge University Press. That publisher had done a good job, but I didn't feel that it would really be geared to the sort of mass market that I wanted to reach. I therefore contacted a literary agent, Al Zuckerman, who had been introduced to me as the brother-in-law of a colleague. I gave him a draft of the first chapter and explained that I wanted it to be the sort of book that would sell in airport bookstores. He told me there was no chance of that. It might sell well to academics and students, but a book like that couldn't break into Jeffrey Archer territory.

I gave Zuckerman a first draft of the book in 1984. He sent it to several publishers and recommended that I accept an offer from Norton, a fairly upmarket American book firm. But I decided instead to take an offer from Bantam Books, a publisher more oriented toward the popular market. Though Bantam had not specialized in publishing science books, its books were widely available in airport bookstores.

Bantam's interest in the book was probably due to one of their editors, Peter Guzzardi. He took his job very seriously and made me rewrite the book so that it would be understandable to non-scientists such as himself. Each time I sent him a rewritten chapter, he sent back a long list of objections and questions he wanted me to clarify. At times I thought the process would never end. But he was right: it is a much better book as a result.

One of the early covers of *A Brief History of Time*

My writing of the book was interrupted by the pneumonia I caught at CERN. It would have been quite impossible to finish the book but for the computer program I was given. It was a bit slow, but then I think slowly, so it suited me quite well. With it I almost completely rewrote my first draft in response to Guzzardi's urgings. I was helped in this revision by one of my students, Brian Whitt.

I had been very impressed by Jacob Bronowski's television series *The Ascent of Man.* (Such a sexist title would not be allowed today.) It gave a feeling for the achievement of the human race in developing from primitive savages only fifteen thousand years ago to our present state. I wanted to convey a similar feeling for our progress toward a complete understanding of the laws that govern the universe. I was sure that nearly everyone is interested in how the universe operates, but most people cannot follow mathematical equations. I don't care much for equations myself. This is partly because it is difficult for me to write them down, but mainly because I don't have an intuitive feeling for equations. Instead, I think in pictorial terms, and my aim in the book was to describe these mental images in words, with the help of familiar analogies and a few diagrams. In this way, I hoped that most people would be able to share in the excitement and feeling of achievement in the remarkable progress that has been made in physics in the last fifty years.

Still, even if I avoided using mathematics, some of the ideas would be difficult to explain. This posed a problem: should I try to explain them and risk people being confused, or should I gloss over the difficulties? Some unfamiliar concepts, such as the fact that observers moving at different velocities measure different time intervals between the same pair of events, were not essential to the picture I wanted to draw. Therefore I felt I could just mention them but not go into depth. But other difficult ideas were essential to what I wanted to get across.

There were two such concepts in particular that I felt I had to include. One was the so-called sum over histories. This is the idea that there is not just a single history for the universe. Rather, there is a collection of every possible history for the universe, and all these histories are equally real (whatever that may mean). The other idea, which is necessary to make mathematical sense of the sum over histories, is that of imaginary time. With hindsight, I now feel that I should have put more effort into explaining these two very difficult concepts, particularly imaginary time, which seems to be the thing in the book with which people have the most trouble. However, it is not really necessary to understand exactly what imaginary time is—just that it is different from what we call real time.

WHEN THE book was nearing publication, a scientist who was sent an advance copy to review for *Nature* magazine was appalled to find it full of errors, with misplaced and erroneously labeled photographs and diagrams. He called Bantam, which was equally appalled and decided that same day to recall and scrap the entire printing. (Copies of the original first edition are now probably quite valuable.) Bantam spent three intense weeks correcting and rechecking the entire book, and it was ready in time to be in bookstores by the April Fools' Day publication date. By then, *Time* magazine had published a profile of me.

Even so, Bantam was taken by surprise by the demand for the book. It was on *The New York Times* bestseller list for 147 weeks and on the London *Times* bestseller list for a record-breaking 237 weeks, has been translated into 40 languages, and has sold over 10 million copies worldwide.

My original title for the book was *From the Big Bang to Black Holes: A Short History of Time,* but Guzzardi turned it around and changed *Short* to *Brief.* It was a stroke of genius and must have contributed to the success of the book. There have been many "brief histories" of this and that since, and even *A Brief History of Thyme.* Imitation is the sincerest form of flattery.

Why did so many people buy it? It is difficult for me to be sure that I'm objective, so I think I will go by what other people said. I found most of the reviews, although favorable, rather unilluminating. They tended to follow a single formula: *Stephen Hawking has Lou Gehrig's disease* (the term used in American reviews) *or motor neurone disease* (in British reviews). *He is confined to a wheelchair, cannot speak, and can only move X number of fingers* (where X seems to vary from one to three, according to which inaccurate article the reviewer read about me). *Yet he has written this book about the biggest question of all: where did we come from and where are we going? The answer Hawking proposes is that the universe is neither created nor destroyed: it just is. In order to formulate this idea, Hawking introduces the concept of imaginary time, which I* (that is, the reviewer) *find a little hard to follow. Still, if Hawking is right and we do find a complete unified theory, we shall really know the mind of God.* (In the proof stage I nearly cut the last sentence in the book, which was that we would know the mind of God. Had I done so, the sales might have been halved.)

Rather more perceptive, I felt, was an article in *The Independent,* a London newspaper, which said that even a serious scientific work such as *A Brief History of Time* could become a cult book. I was rather flattered to have my book compared to *Zen and the Art of Motorcycle Maintenance.* I hope that, like *Zen,* it gives people the feeling that they need not be cut off from the great intellectual and philosophical questions.

Undoubtedly, the human interest story of how I have managed to be a theoretical physicist despite my disability has helped. But those who bought the book because of the human interest angle may have been disappointed, because it contains only a couple of references to my condition. The book was intended as a history of the universe, not of me. This has not prevented accusations that Bantam shamefully exploited my illness and that I cooperated with this by allowing my picture to appear on the cover. In fact, under my contract I had no control over the cover. I did, however, manage to persuade the publisher to use a better photograph on the British edition than the miserable and out-of-date photo used on the American edition. Bantam will not change the photo on the American cover, however, because it says that the American public now identifies that photo with the book.

It has also been suggested that many people bought the book to display on the bookcase or on the coffee table, without having actually read it. I am sure this happens, though I don't know that it is any more so than with most other serious books. I do know that at least some people must have waded into it, because each day I get a pile of letters about that book, many asking questions or making detailed comments that indicate that they have read it, even if they do not understand all of it. I also get stopped by strangers on the street who tell me how much they enjoyed it. The frequency with which I receive such public congratulations (though of course I am more distinctive, if not more distinguished, than most authors) seems to indicate that at least a proportion of those who buy the book actually do read it.

Since *A Brief History of Time,* I have written other books to explain science to the wider public: *Black Holes, and Baby Universes, The Universe in a Nutshell,* and *The Grand Design.* I think it is important that people have a basic understanding of science so they can make informed decisions in an increasingly scientific and technological world. My daughter, Lucy, and I have also written a series of "George" books, which are scientifically based adventure stories for children, the adults of tomorrow.

11

TIME TRAVEL

I N 1990 KIP THORNE SUGGESTED THAT IT MIGHT BE
possible to travel into the past by going through wormholes. I therefore thought it would be worthwhile to in-
vestigate whether time travel is allowed by the laws of
physics.

To speculate openly about time travel is tricky for several reasons. If the press picked up that the government
was funding research into time travel, there would be either an outcry at the waste of public money or a demand
that the research be classified for military purposes. After
all, how could we protect ourselves if the Russians or Chi-

nese had time travel and we didn't? They could bring back Comrades Stalin and Mao. In physics circles, there are only a few of us foolhardy enough to work on a subject that some consider unserious and politically incorrect. So we disguise our focus by using technical terms, such as "particle histories that are closed," that are code for time travel.

THE FIRST scientific description of time was given in 1689 by Sir Isaac Newton, who held the Lucasian chair at Cambridge that I used to occupy (though it wasn't electrically operated in his time). In Newton's theory, time was absolute and marched on relentlessly. There was no turning back and returning to an earlier age. The situation changed, however, when Einstein formulated his general theory of relativity, in which space-time was curved and distorted by the matter and energy in the universe. Time still increased locally, but there was now the possibility that space-time could be warped so much that one could move on a path that would bring one back before one set out.

One possibility that would allow for this would be wormholes, hypothetical tubes of space-time that might connect different regions of space and time. The idea is that you step into one mouth of the wormhole and step out of the other in a different place and at a different time. Wormholes, if they exist, would be ideal for rapid space travel. You might go through a wormhole to the other side of the galaxy and be back in time for dinner. However, one can show that if wormholes exist, you could also use them to get back before you set out. One would then think that you could do something like blow up your own spaceship on its original launch pad to prevent you from setting out in the first place. This is a variation of the so-called grandfather paradox: What happens if you go back in time and kill your grandfather before your father was conceived? Would you then exist in the current present? If not, you wouldn't exist to go back and kill your grandfather. Of course, this is a paradox only if you believe you have the free will to do what you like and change history when you go back in time.

The real question is whether the laws of physics allow wormholes and space-time to be so warped that a macroscopic body such as a spaceship can return to its own past. According to Einstein's theory, a spaceship necessarily travels at less than the local speed of light, and follows what is called a "time-like path" through space-time. Thus one can formulate the question in technical terms: does space-time admit time-like curves that are closed—that is, time-like curves that return to their starting point again and again?

There are three levels on which we can try to answer this question. The first is Einstein's general theory of relativity. This is what is called a classical theory, which is to say it assumes the universe has a well-defined history, without any uncertainty. For classical general relativity, we have a fairly complete picture of how time travel might work. We know, however, that classical theory can't be quite right, because we observe that matter in the universe is subject to fluctuations, and its behavior cannot be predicted precisely.

In the 1920s a new paradigm called quantum theory was developed to describe these fluctuations and quantify the uncertainty. One can therefore ask the question about time travel on this second level, called the semiclassical theory. In this, one considers quantum matter fields against a classical space-time background. Here the picture is less complete, but at least we have some idea how to proceed.

Finally, one has the full quantum theory of gravity, whatever that may be. Here it is not clear even how to pose the question "Is time travel possible?" Maybe the best one can do is to ask how observers at infinity would interpret their measurements. Would they think that time travel had taken place in the interior of the space-time?

RETURNING TO the classical theory: flat space-time does not contain closed time-like curves. Nor do other solutions of the Einstein equations that were known early on. It was therefore a great shock to Einstein when in 1949 Kurt Gödel discovered a solution that represented a universe full of rotating matter, with closed time-like curves through every point. The Gödel solution required a cosmological constant, which is known to exist, though other solutions were subsequently found without one.

A particularly interesting case to illustrate this would be two cosmic strings moving at high speed past each other. As their name suggests, cosmic strings are objects with length but a tiny cross section. Some theories of elementary particles predict their occurrence. The gravitational field of a single cosmic string is flat space with a wedge cut out, with the string at its sharp end. Thus if one goes in a circle around a cosmic string, the distance in space is less than one would expect, but time is not affected. This means that the space-time around a single cosmic string does not contain any closed time-like curves.

However, if there is a second cosmic string moving with respect to the first, the wedge that is cut out for it will shorten both spatial distances and time intervals. If the cosmic strings are moving at nearly the speed of light relative to each other, the saving of time going around both strings can be so great that one arrives back before one set out. In other words, there are closed time-like curves that one can follow to travel into the past.

The cosmic string space-time contains matter that

has positive energy density, and thus it is physically reasonable. However, the warping that produces the closed time-like curves extends all the way out to infinity and back to the infinite past. Thus these space-times were created with time travel in them. We have no reason to believe that our own universe was created in such a warped fashion, and we have no reliable evidence of visitors from the future. (Discounting, of course, the conspiracy theory that UFOs are from the future, which the government knows and is covering up. But governments' record of cover-ups is not that good.) One should therefore assume that there are no closed time-like curves to the past of some surface of constant time S.

The question is then whether some advanced civilization could build a time machine. That is, could it modify the space-time to the future of S, so that closed time-like curves appeared in a finite region? I say "a finite region" because no matter how advanced the civilization becomes, it could presumably control only a finite part of the universe.

In science, finding the right formulation of a problem is often the key to solving it, and this was a good example. To define what was meant by a finite time machine, I went back to some early work of mine. I defined the future Cauchy development of S to be the set of points of space-time where events are determined completely by what happened on S. In other words, it is the region of space-time where every possible path that moves at less than the speed of light comes from S. However, if an advanced civilization managed to build a time machine, there would be a closed time-like curve, C, to the future of S. C will go round and round in the future of S, but it will not go back and intersect S. This means that points on C will not lie in the Cauchy development of S. Thus S will have a Cauchy horizon, a surface that is a future boundary to the Cauchy development of S.

Cauchy horizons occur inside some black hole solutions, or in anti–de Sitter space. However, in these cases, the light rays that form the Cauchy horizon start at infinity or at singularities. To create such a Cauchy horizon would require either warping space-time all the way out to infinity or the occurrence of a singularity in space-time. Warping space-time all the way to infinity would theoretically be beyond the powers of even the most advanced civilization, which could warp space-time only in a finite region. The advanced civilization could assemble enough matter to cause a gravitational collapse, which would produce a space-time singularity, at least according to classical general relativity. But the Einstein equations could not be defined at the singularity, so one could not predict what would happen beyond the Cauchy horizon, and in particular whether there would be any closed time-like curves.

One should therefore take as the criterion for a time

With Roger Penrose (top, middle) and Kip Thorne (bottom, far left), among others (*above*). With Roger and his wife, Vanessa (*below*).

machine what I call a finitely generated Cauchy horizon. That is a Cauchy horizon generated by light rays that all emerge from a compact region. In other words, they don't come in from infinity, or from a singularity, but originate from a finite region containing closed time-like curves, the sort of region we have supposed our advanced civilization would create.

Adopting this definition as the footprint of a time machine has the advantage that one can use the machinery of causal structure that Roger Penrose and I developed to study singularities and black holes. Even without using the Einstein equations, I was able to show that, in general, a finitely generated Cauchy horizon will contain a closed light ray, or a light ray that keeps coming back to the same point over and over again. Moreover, each time the light comes around, it will be more and more blue-shifted, so the images will get bluer and bluer. The light rays may get defocused sufficiently each time round so that the energy of light doesn't build up and become infinite. However, the blue shift will mean that a particle of light will have only a finite history, as defined by its own measure of time, even though it goes round and round in a finite region and does not hit a curvature singularity.

One might not care if a particle of light completes its history in a finite time. But I was also able to prove that there would be paths moving at less than the speed of light that had only finite duration. These could be the histories of observers who would be trapped in a finite region before the Cauchy horizon and would go round and round faster and faster until they reached the speed of light in a finite time.

So if a beautiful alien in a flying saucer invites you into her time machine, step with care. You might fall into one of these trapped repeating histories of only finite duration.

AS I said, these results depend not on the Einstein equations but only on the way space-time would have to warp to produce closed time-like curves in a finite region. However, one can now ask: What kind of matter would an advanced civilization need in order to warp space-time so as to build a finite-sized time machine? Can it have positive energy density everywhere, like in the cosmic string space-time? One might imagine that one could build a finite time machine using finite loops of cosmic string and have the energy density positive everywhere. I'm sorry to disappoint people wanting to return to the past, but it can't be done with positive energy density everywhere. I proved that to build a finite time machine, you need negative energy.

In classical theory, all physically reasonable fields obey the weak energy condition, which says that the energy density for any observer is greater than or equal to zero. Thus time machines of finite size are ruled out in the purely classical theory. However, the situation is different in the semi-classical theory, in which one considers quantum fields on a classical space-time background. The uncertainty principle of quantum theory means that fields are always fluctuating up and down, even in apparently empty space. These quantum fluctuations make the energy density infinite. Thus one has to subtract an infinite quantity to get the finite energy density that is observed. Otherwise, the energy density would curve space-time up into a single point. This subtraction can leave the expectation value of the energy negative, at least locally. Even in flat space, one can find quantum states in which the expectation value of the energy density is negative locally, although the integrated total energy is positive.

One might wonder whether these negative expectation values actually cause space-time to warp in the appropriate way. But it seems they must. The uncertainty principle of quantum theory allows particles and radiation to leak out of a black hole. This causes the black hole to lose mass, thus evaporating slowly. For the horizon of the black hole to shrink in size, the energy density on the horizon must be negative and warp space-time to make light rays diverge from each other. If the energy density were always positive and warped space-time so as to bend light rays toward each other, the area of the horizon of a black hole could only increase with time.

The evaporation of black holes shows that the quantum energy momentum tensor of matter can sometimes warp space-time in the direction that would be needed to build a time machine. One might imagine, therefore, that some very advanced civilization could arrange that the expectation value of the energy density would be sufficiently negative to form a time machine that could be used by macroscopic objects.

But there's an important difference between a black hole horizon and the horizon in a time machine, which contains closed light rays that keep going round and round. This would make the energy density infinite, which would mean that a person or a spaceship that tried to cross the horizon to get into the time machine would get wiped out by a bolt of radiation. This might be a warning from nature not to meddle with the past.

So the future looks black for time travel, or should I say blindingly white? However, the expectation value of the energy momentum tensor depends on the quantum state of the fields on the background. One might speculate that there could be quantum states where the energy density was finite on the horizon, and there are examples where this is the case. How you achieve such a quantum state, or whether it would be stable against objects crossing the horizon, we don't know. But it might be within the capabilities of an advanced civilization.

This is a question that physicists should be free to discuss without being laughed at or scorned. Even if it turns out that time travel is impossible, it is important that we understand *why* it is impossible.

We don't know much about the fully quantized theory of gravity. However, one might expect it to differ from the semi-classical theory only on the Planck length, a million billion billion billionth part of a centimeter. Quantum fluctuations on the background of space-time may well create wormholes and time travel on a microscopic scale, but according to the general theory of relativity, macroscopic bodies will not be able to return to their past.

Even if some different theory is discovered in the future, I don't think time travel will ever be possible. If it were, we would have been overrun by tourists from the future by now.

12

IMAGINARY TIME

WHILE WE WERE AT CALTECH, WE VISITED SANTA Barbara, which is a two-hour drive up the coast. There I worked with my friend and collaborator Jim Hartle on a new way of calculating how particles would be emitted by a black hole, adding up all the possible paths the particle could take to escape from the hole. We found that the probability that a particle would be emitted by a black hole was related to the probability that a particle would fall into the hole, in the same way that the probabilities for emission and absorption were related for a hot body. This again showed that black holes behave as if they have a temperature and an entropy proportional to their horizon area.

Our calculation made use of the concept of imaginary time, which can be regarded as a direction of time at right angles to ordinary real time. When I returned to Cambridge I developed this idea further with two of my former research students, Gary Gibbons and Malcolm Perry. We replaced ordinary time with imaginary time. This is called the Euclidean approach, because it makes time become a fourth direction of space. It met with a lot of resistance at first but is now generally accepted as the best way to study quantum gravity. The Euclidean space of black hole time is smooth and contains no singularity at which the equations of physics would break down. It solved the fundamental problem that the singularity theorems of Penrose and I had raised: that predictability would break down because of the singularity. Using the Euclidean approach, we were able to understand the deep reasons why black holes behaved like hot bodies and had entropy. Gary and I also showed that a universe that was expanding at an ever-increasing rate would behave as if it had an effective temperature like that of a black hole. At the time we thought this temperature could never be observed, but its significance became apparent fourteen years later.

With Don Page (top, far left), Kip Thorne (bottom, third from left), and
Jim Hartle (bottom, far right), among others

I HAD been working mainly on black holes, but my
interest in cosmology was renewed by the suggestion
that the early universe had gone through a period of inflationary expansion. Its size would have grown at an
ever-increasing rate, just as prices go up in the shops. In
1982, using Euclidean methods, I showed that such a
universe would become slightly non-uniform. Similar results were obtained by the Russian scientist Viatcheslav
Mukhanov about the same time, but that only became
known later in the West.

These non-uniformities can be regarded as arising

from thermal fluctuations due to the effective temperature in an inflationary universe that Gary Gibbons and I had discovered eight years earlier. Several other people later made similar predictions. I held a workshop in Cambridge, attended by all the major players in the field, and at this meeting we established most of our present picture of inflation, including the all-important density fluctuations that give rise to galaxy formation, and so to our existence.

This was ten years before the Cosmic Background Explorer (COBE) satellite recorded differences in the microwave background in different directions produced by the density fluctuations. So again, in the study of gravity, theory was ahead of experiment. These fluctuations were later confirmed by the Wilkinson Microwave Anisotropy Probe (WMAP) and the Planck satellite, and were found to agree exactly with predictions.

THE ORIGINAL scenario for inflation was that the universe began with a Big Bang singularity. As the universe expanded, it was supposed somehow to get into an inflationary state. I thought this was an unsatisfactory explanation, because all equations would break down at a singularity, as previously discussed. But unless one knew what came out of the initial singularity, one could not calculate how the universe would develop. Cosmology would not have any predictive power. What was needed was a space-time without a singularity, like in the Euclidean version of a black hole.

AFTER THE workshop in Cambridge, I spent the summer at the Institute for Theoretical Physics, Santa Barbara, which had just been set up. I talked to Jim Hartle about how to apply the Euclidean approach to cosmology. According to the Euclidean approach, the quantum behavior of the universe is given by a Feynman sum over a certain class of histories in imaginary time. Because imaginary time behaves like another direction in space, histories in imaginary time can be closed surfaces, like the surface of the Earth, with no beginning or end.

Jim and I decided that this was the most natural choice of class, indeed the only natural choice. We formulated the no-boundary proposal: that the boundary condition of the universe is that it is closed without boundary. According to the no-boundary proposal, the beginning of the universe was like the South Pole of the Earth, with degrees of latitude playing the role of imaginary time. The universe would start as a point at the South Pole. As one moves north, the circles of constant latitude, representing the size of the universe, would expand. To ask what happened before the beginning of the universe would thus become a meaningless question, because there is nothing south of the South Pole.

Time, as measured in degrees of latitude, would have a beginning at the South Pole, but the South Pole is much like any other point on the globe. The same laws of nature hold at the South Pole as in other places. This would remove the age-old objection to the universe having a beginning—that it would be a place where the normal laws broke down. The beginning of the universe would instead be governed by the laws of science. We had sidestepped the scientific and philosophical difficulty of time having a beginning by turning it into a direction in space.

The no-boundary condition implies that the universe will be spontaneously created out of nothing. At first it seemed that the no-boundary proposal did not predict enough inflation, but I later realized that the probability of a given configuration of the universe has to be weighted by the volume of the configuration. Recently Jim Hartle, Thomas Hertog (another former student), and I have discovered that there is a duality between inflating universes and spaces that have negative curvature. This allows us to formulate the no-boundary proposal in a new way and to use the considerable technical machinery that has been developed for such spaces. The no-boundary proposal predicts that the universe will start out almost completely smooth, but with tiny departures. These will grow as the universe expands, and will lead to the formation of galaxies, stars, and all the other structures in the universe, including living beings. The no-boundary condition is the key to creation, the reason we are here.

13

NO BOUNDARIES

WHEN I WAS TWENTY-ONE AND CONTRACTED ALS, I felt it was very unfair. Why should this happen to me? At the time, I thought my life was over and that I would never realize the potential I felt I had. But now, fifty years later, I can be quietly satisfied with my life. I have been married twice and have three beautiful and accomplished children. I have been successful in my scientific career: I think most theoretical physicists would agree that my prediction of quantum emission from black holes is correct, though it has not so far earned me a Nobel Prize because it is very difficult to verify experimentally. On the other hand, I won the even more valuable Fundamental Physics Prize, awarded for the theoretical significance of the discovery despite the fact that it has not been confirmed by experiment.

My disability has not been a serious handicap in my scientific work. In fact, in some ways I guess it has been an asset: I haven't had to lecture or teach undergraduates, and I haven't had to sit on tedious and time-consuming committees. So I have been able to devote myself completely to research.

To my colleagues, I'm just another physicist, but to the wider public, I became possibly the best-known scientist in the world. This is partly because scientists, apart from Einstein, are not widely known rock stars, and partly because I fit the stereotype of a disabled genius. I can't disguise myself with a wig and dark glasses—the wheelchair gives me away.

Being well known and easily recognizable has its pluses and minuses. The minuses are that it can be difficult to do ordinary things such as shopping without being besieged by people wanting photographs, and that in the past the press has taken an unhealthy interest in my private life. But the minuses are more than outweighed by the pluses. People seem genuinely pleased to see me. I even had my biggest-ever audience when I was the anchor for the Paralympic Games in London in 2012.

I have had a full and satisfying life. I believe that dis-

Anchoring the Paralympic Games in 2012

abled people should concentrate on things that their handicap doesn't prevent them from doing and not regret those they can't do. In my case, I have managed to do most things I wanted. I have traveled widely. I visited the Soviet Union seven times. The first time I went with a student party in which one member, a Baptist, wished to distribute Russian-language Bibles and asked us to smuggle them in. We managed this undetected, but by the time we were on our way out the authorities had discovered what we had done and detained us for a while. However, to charge us with smuggling Bibles would have

Visiting the Temple of Heaven in Beijing

caused an international incident and unfavorable pub-
licity, so they let us go after a few hours. The other six
visits were to see Russian scientists who at the time were
not allowed to travel to the West. After the end of the Soviet Union in 1990, many of the best scientists left for the
West, so I have not been to Russia since then.

I have also visited Japan six times, China three times, and every continent, including Antarctica, with the exception of Australia. I have met the presidents of South Korea, China, India, Ireland, Chile, and the United States. I have lectured in the Great Hall of the People in Beijing and in the White House. I have been under the sea in a submarine and up in a hot air balloon and a zero-gravity flight, and I'm booked to go into space with Virgin Galactic.

My early work showed that classical general relativity broke down at singularities in the Big Bang and black

Meeting Queen Elizabeth II with my daughter, Lucy

Experiencing zero gravity

holes. My later work has shown how quantum theory can predict what happens at the beginning and end of time. It has been a glorious time to be alive and doing research in theoretical physics. I'm happy if I have added something to our understanding of the universe.

ILLUSTRATION CREDITS

Courtesy of Mary Hawking: pages iv, 2, 4, 5, 8, 10, 12, 15, 18 (*bottom*), 25, and 27

Courtesy of Stephen Hawking: 7, 18 (*top*), 20, 21 (*bottom*), 39, 54, 62, 74, 77, 84, 88, 90, 91, and 108 (*bottom*)

National Archives and Records Administration: page 11

Herts Advertiser: page 28

Gillman & Soame: pages 32, 33, 34–5, 38

Suzanne McClenahan: page 49

Lafayette Photography: page 51

John McClenahan: page 55

Courtesy of the Archives, California Institute of Technology: pages 78 and 79

Bernard Carr: pages 108 (*top*) and 116

Judith Croasdell: page 123

Zhang Chao Wu: page 124

Alpha/Globe Photos, Inc.: page 125

Steve Boxall: page 126

ABOUT THE AUTHOR

STEPHEN HAWKING was the Lucasian Professor of Mathematics at the University of Cambridge for thirty years and has been the recipient of numerous awards and honors including, most recently, the Presidential Medal of Freedom. His books for the general reader include the classic *A Brief History of Time,* the essay collection *Black Holes and Baby Universes, The Universe in a Nutshell,* and, with Leonard Mlodinow, *A Briefer History of Time* and *The Grand Design.*

ABOUT THE TYPE

This book was set in Stone, a typeface created in 1988 by the type and graphic designer Sumner Stone (b. 1945). This typeface was designed to satisfy the requirements of low-resolution laser printing. Its traditional design blends harmoniously with many typefaces, making it appropriate for a variety of applications.